RAILWA· TIMES ·

Contents

The Transport Treasury

·TIMES· SERIES

Front Cover: An immaculately turned out 'Coronation', No 46232 *Duchess of Montrose,* heads the carmine and cream coaches of the up 'Royal Scot' out of Carlisle on 20 August 1952. At this date these Polmadie Pacifics were working south from Carlisle although in later years locomotive changes would take place here. The south end of the impressive train shed seen in this view was to be swept away in 1957/8 when extensive rebuilding of the station occurred. Entering service at Camden in 1938, No 46232 would later be transferred to the Glasgow shed from where it would be withdrawn at the end of 1962. *David P Williams colour archive*

Above: Ex-GWR railcar No W20W, seen here at Worcester on 2 April 1965, ventured onto pastures new when in the summer of 1952 it journeyed north to the Eastern and North Eastern regions for trials. *RIC 20048*

Opposite: Marking the centenary of the completion of the GNR main line and the opening of King's Cross station, a special train was run between London and York and return on 28 September headed by Class A4 No 60007 *Sir Nigel Gresley* carrying the 'Centenaries Express' headboard and seen here light engine at York. *OTA201005*

Rear Cover: Prelude to the almost wholesale destruction of rail services on the Isle of Wight is this view dating from 16 April 1949 on the route between Merstone and Ventnor West which saw its passenger service withdrawn in September 1952. Class A1X No 8 *Freshwater*, with recently applied 'British Railways' lettering on its tank sides, has just a single coach in tow as it leaves St. Lawrence station, the penultimate stop on the branch. Along with its classmate W13 *Carisbrooke* it was transferred back to the mainland in May 1949 and renumbered 32646. After a stint as a pub sign, 'The Hayling Billy', it can now be seen back on the Isle of Wight in its old guise as W8. *LSPMA 1362*

Copies of many of the images within **RAILWAY TIMES** are available for purchase / download.
Wherever possible illustrative material has been chosen dating from 1952 or is used to reflect the content described.

In addition the Transport Treasury Archive contains tens of thousands of other UK, Irish
and some European railway photographs.

© Jeffery Grayer. Images (unless credited otherwise) and Design: The Transport Treasury 2025

ISBN 978-1-913251-90-1

First published in 2025 by Transport Treasury Publishing Ltd.,
16 Highworth Close, High Wycombe, HP13 7PJ

www.ttpublishing.co.uk or for editorial issues and contributions, email to **admin@ttpublishing.com**

Printed in the Malta by the Gutenberg Press.

Introduction

Welcome to 'Railway Times' No 5 covering 1952, a year which was overshadowed by the tragic accident at Harrow and Wealdstone in which over 100 people lost their lives and 340 were injured, not to mention the writing off of two locomotives and numerous items of coaching stock. One of the locomotives lost was the recently rebuilt Princess Royal class Pacific No 46202 newly named *Princess Anne* which is the subject of one of the features within. Continuing the rather mournful theme we cover the arrangements for King George VI's funeral train.

Also on a rather negative note we examine the role of the Branch Line Committee set up to review the finances of the nation's branch lines whose deliberations had led to the closing of many unprofitable routes in the five years since Nationalisation.

Some of the 1952 closures are featured here in more detail including the Stratford-upon-Avon & Midland Junction line, the Portland branch, the Mid Suffolk and the first of what was to be the eventual decimation of the Isle of Wight lines, that serving Ventnor West.

More positively, the year also saw trials of a new railbus concept in the shape of the ACV unit somewhat disparagingly, but understandably, nicknamed the 'Flying Brick' together with trials of an ex-GWR railcar well outside its normal operating area in Yorkshire and Lincolnshire. BR's standard designs continued to roll off the production lines and two are featured, namely the Clan Pacifics and the Class 3 2-6-2Ts. In connection with the new Standard designs some background is provided on the characters behind the Franco-Crosti boilers that would be fitted to some of the Standard 9Fs in the years to come.

1952 was a year of centenary celebrations marking the opening of King's Cross station, the completion of the GNR main line between Peterborough and Retford and 100 years of Brighton Works. One of the regular features of 'Railway Times' is Brief Encounters which again includes some of the more quirky and bizarre happenings on BR. As ever, I hope you find something of interest within these pages.

Editor : Jeffery Grayer

Precursor to catastrophe - the Turbomotive rebuild

The first pair of what would become known as the Princess Royal class of Pacifics emerged from Crewe Works in 1933. Following a visit to Sweden to evaluate the performance of a steam turbine locomotive, Stanier decided rather radically that the third Pacific in the series would be a geared turbine locomotive that became known as the Turbomotive. It was constructed in 1935 the same year that the remainder of the class, in conventional form, were also completed. It however remained nameless until reconstruction in 1952.

Deemed a success, No 6202 spent much of its existence operating heavy express trains between Euston and Liverpool Lime Street. Soon after the outbreak of WW2 it was stored for a time before reinstatement in 1942. The advent of a turbine failure under BR ownership in 1949 meant that repair costs were considered uneconomic given the straitened financial circumstances of the post-war period. Once again the locomotive was taken out of service pending a rebuild into a more conventional machine. Little did we know that the rebuilt locomotive would meet such a tragic end just a few weeks after emerging from the works as *Princess Anne* complete with new mainframes and a revised cylinder arrangement.

On that fateful morning of the Harrow disaster on 8 October 1952 it had run only 11,443 miles in the two months since rebuilding had been completed on 15 August 1952. It was so badly damaged in the accident that it would never run again although its boiler was repaired and reused as was the tender which ended up paired to 8F No 48134. The sad remains of No 46202 were not officially withdrawn until May 1954 having spent some time languishing outside Crewe Works.

Above: This undated view of the Turbomotive No 46202 was taken at Crewe Works. *NS203427 / Transport Treasury*

Opposite top: No 46202 in the process of rebuilding. The reverse turbine at least has been removed - along with the front driving wheels on to which the turbine drive was attached. Alongside is No 3233, a former L&NWR 'Special' 0-6-0ST 2F used as a Works shunter. It was withdrawn in 1954, the same year the remains of No 46202 were finally condemned.

Opposite bottom: This view of the rebuilt Princess Royal class No 46202 *Princess Anne* was taken at Shrewsbury on 25 August 1952 just six weeks before the tragedy of Harrow was to unfold. *NS203621 / Transport Treasury*

Railway staff survey the scene of carnage at Harrow & Wealdstone. *REV 73C 6-6 / Transport Treasury*

The mangled wreck of No 46202 *Princess Anne,* now reduced to an 0-6-2, occupies a siding at Harrow & Wealdstone.

A crane prepares to lift some more of the wreckage whilst the boiler of the Pacific, together with one pair of driving wheels, have already been removed from the chassis. *REV 73C 6-3 / Transport Treasury*

The gaunt remains of the recently converted Pacific, still bearing its nameplate, are testament to the force of the impact. *REV 73C 6-4 / Transport Treasury*

A pair of driving wheels is lifted ready to be deposited in one of the adjacent wagons. *REV 73C 6-5 / Transport Treasury*

8 October 1952 was a bad day for the railways when an horrific accident occurred at Harrow & Wealdstone station resulting in the loss of 112 lives, a total which involved 37 staff of the LMR including 3 locomotive crew members with 340 injured. In essence the 8:15pm Perth to Euston express headed by No 46242 *City of Glasgow,* carrying some 90 passengers and running some 80 minutes late due to fog, ran into the rear of the 7:31 am local train from Tring to Euston headed by Fowler 2-6-4T No 42389 which was carrying about 800 passengers, and was stationary at the up fast platform. Matters were compounded when the 8am Euston to Liverpool and Manchester express travelling on the down fast line with 186 passengers on board collided with the wreckage very soon after, thus precluding the taking of any preventative measures to alert the oncoming express to the danger.

The locomotives heading the down train, pilot Jubilee No 45637 *Windward Islands* and train engine No 46202

Princess Anne together with the leading coaches of the train mounted the platform, blocking the up electrified line. Preliminary enquiries established that for some unknown reason the driver of the train from Perth had ignored a distant signal at caution and two stop signals at danger with the Chief Inspecting Officer stating that he doubted whether the circumstances in which the signals were passed would ever be satisfactorily explained as the driver and fireman had both been killed.

The Perth locomotive was completely buried by the wreckage and the station footbridge was badly damaged whilst the rear three coaches of the Tring service were virtually demolished although the remaining six coaches remained on the rails. In total 14 coaches were destroyed and three locomotives severely damaged such that Nos. 46202 and 45637 were subsequently scrapped, the latter being the first of the Jubilee class to suffer this fate. On 9 November the remains of No 45637, the boiler of No 46202 and tenders from both locomotives were loaded into wagons and taken to Crewe.

It had been decided that the chassis of No 46202 was capable of being removed by rail with the installation of a temporary bogie and trailing truck. The initial move was only as far as Willesden but on 13 November the remains of No 46202 were noted proceeding northwards on the down slow line behind an 8F 2-8-0. The sad remnants of the Princess Royal and of the Jubilee were later observed at Crewe Works. No 46242 was repaired, remaining in service until 1963, but the other two locomotives were so badly damaged that they were written off. The Harrow accident remains the UK's second worst rail disaster after the Quintinshill accident of 1915. The slow lines through Harrow station reopened early the following day whilst the electrified lines, which had been used by cranes to remove the wreckage, reopened in the early hours of 11 October. The fast lines were reopened, with a speed restriction, in the evening of 12 October and a temporary footbridge was provided the same evening. One significant legacy of the disaster was the accelerated introduction of the Automatic Warning System (AWS) alerting a driver when he had passed a signal at danger. A memorial plaque was unveiled in 2002 to mark the 50th anniversary of the disaster.

Class Y4 No 68129 takes a break between shunting duties at Stratford in September 1952 at the end of which month it became No 33 in the Departmental series, remaining in traffic until the end of 1963. *AF 0491 / Transport Treasury*

Bridgwater is the location of this 1952 image showing 0-6-0ST No 2194 *Kidwelly* well away from its former home on the Burry Port & Gwendraeth Valley Railway which had been absorbed by the GWR at the Grouping. Based at nearby Taunton shed, it had been found very suitable for shunting the lines in Bridgwater Docks but this did not ensure its longevity after nationalisation, being withdrawn in February the year following this photograph. *NS 200139 / Transport Treasury*

Opposite top: The exceptional number of spectators at the rural byway of Hole must surely be explained by the term 'enthusiasts' for the line from Halwill to Torrington barely saw more than one or two passengers per train normally. Here Class E1/R No 32608 waits with its single coach for the return of its contingent. *NS 200785 / Transport Treasury*

Opposite bottom: Passing the extensive yards, Class J27 0-6-0 No 65796 rounds the curve at Percy Main with a freight service on an unrecorded date in 1952. By the time of its withdrawal in 1966 this old workhorse had put in close to 60 years service. Percy Main was named after a colliery owned by the Percy family and was later absorbed into North Shields. *NS 201779 / Transport Treasury*

This page, top: Darkening the sky at Chester Northgate in 1952 is the now preserved Class D11 'Large Director' No 62660 *Butler-Henderson* whilst at the other platform is Class C13 4-4-2T No 67430. Northgate station, the former western terminus of the Cheshire Lines Committee line from Manchester, closed in 1969. *NS 202571 / Transport Treasury*

This page, centre: Before the former route from Ashchurch to Great Malvern was cut back to Upton-on-Severn in 1952, this view dating from earlier that year shows Johnson 0-4-4T No 58076 with its single coach, telling you all you need to know about the profitability of this route, at the intermediate station of Malvern (Hanley Road). Originally known as Malvern Wells it changed its name to Hanley Road in March 1951, only to close in December the following year. *NS 203090A / Transport Treasury*

This page, bottom: With 'British Railways' emblazoned on its tank sides, McIntosh 0-4-4T No 55135 is on station pilot duty at Glasgow's St. Enoch station in 1952. It would be condemned in July the following year. *NS 204258A / Transport Treasury*

BRITISH RAILWAYS

TIME TABLE

OF

ROYAL TRAIN

FROM

PADDINGTON

TO

WINDSOR & ETON (CENTRAL)

ON

FRIDAY, 15th FEBRUARY, 1952

In connection with the Funeral of

HIS LATE MAJESTY

KING GEORGE VI

Paddington - - - depart 12.35 p.m.

Windsor & Eton (Central) arrive 1.10 p.m.

PRIVATE.—For use of the Staff concerned only.   NOTICE No. 10.

BRITISH RAILWAYS

(WESTERN OPERATING AREA)

NOTICE

OF

ROYAL TRAIN

FROM

PADDINGTON

TO

WINDSOR & ETON

(CENTRAL)

AND RETURN

In connection with the Funeral of

HIS LATE MAJESTY

KING GEORGE VI

ON

FRIDAY, 15th FEBRUARY, 1952

This Notice, which will be distributed by the District Operating Superintendent, Paddington, to all Staff affected in the London District, must be acknowledged to the District Operating Superintendent immediately on receipt by *Telegram* as follows : " ARNO GROVE TEN."

A King passes

The whole nation was in mourning following the death of King George VI on 6 February 1952. His final journey from Sandringham to Windsor was to be accomplished by special funeral trains organised by the Eastern and Western regions of BR.

Three days before the day of the funeral the King's body had been brought from Sandringham to Wolferton station on the Hunstanton line, the coffin having been placed onto a gun carriage of the King's Troop Royal Artillery. The King's brother the Duke of Gloucester and his son-in-law the Duke of Edinburgh followed on foot to the station whilst Princess Elizabeth and her sister Princess Margaret together with the Queen Mother followed by car. The Sandringham estate staff and their families followed in procession part of the way with the general public lining much of the route. Upon arrival at Wolferton the coffin was removed from the gun carriage by eight soldiers of the Grenadier Guards and placed into the carriage which had previously borne the remains of George V. The train pulled out of Wolferton at 12:05pm headed by Class B2 No 61617 *Ford Castle* rather than the traditional royal train locomotive No 61671 *Royal Sovereign* which at the time was in the Works for attention. Eleven minutes were allowed for the run to King's Lynn where a reversal was necessary, this being accomplished in the allotted four minute turnround.

The onward journey to King's Cross was in the hands of No 70000 *Britannia* hauling the nine coach train. The cab roofs of the locomotives were painted white for the occasion, this being a tradition of royal train locomotives. *Britannia* arrived at King's Cross on time with the route's signalmen having been duly warned of dire consequences should the path be impeded. Special instructions were issued that trains passing the funeral train should reduce their speed to 15mph, all shunting in stations and yards should be paused whilst the train passed, the public should be excluded from stations whilst the specials went by and that where possible quietness should reign for the passage of the special. On arrival in London the King's body was taken to Westminster Hall where it lay in state. This was perhaps one of the first occasions, if not in fact the first, that a Britannia Pacific had appeared at King's Cross. The fact that the Pacific had a larger water capacity than a B2 may have been a factor in its participation as due to pw works the water troughs at Langley were out of use at the time.

The second part of his final journey involved the WR who organised the train from Paddington to Windsor on 15 February. The GWR predecessors of the WR had operated a similar funeral train in January 1936 for King George V and were well versed in the etiquette. Back in 1936 the train had been hauled by Castle class No 4082 appropriately enough named *Windsor Castle*. Although it had been intended to roster the same locomotive to King George VI's train, having had a heavy overhaul a year before, upon inspection a week before it was considered that there was too much work needed to get it into the suitable condition that such an occasion demanded. So it was decided that a newer model should be used and No 7013 *Bristol Castle*, then just 2½ years old and which had been overhauled three months previously, stepped up to the task. In a nod to history it was decided to swap the identities of the two locomotives thus No 7013 became No 4082 and vice versa. Nameplates, numberplates and the plaque marking the 1924 Royal visit to Swindon Works were also swapped and this changeover became permanent until both locomotives were withdrawn in the 1960s.

On the day of the funeral a gun carriage procession from Westminster to Paddington delivered the body to the area between Paddington's platforms 8 and 9 and various appropriate adornments were added to the station. Six special trains from Paddington to Windsor ran on this occasion with the funeral train departing at 12:35pm, reaching Windsor at 1:10pm. The royal family boarded the same train with other guests following in the remaining specials. As a mark of respect the RAF was grounded during the time of the funeral and on one transatlantic flight from London to New York, passing over Windsor at the time of the funeral, all of the passengers rose from their seats and bowed their heads. Thus successfully ended a meticulously choreographed occasion which involved substitutions of locomotives on both legs of the sovereign's final journey. However, apparently the WR authorities were a little peeved that their loco swapping subterfuge had leaked out and become generally known at least amongst the loco spotting fraternity.

Opposite top: No 7013 *Bristol Castle* masquerading as No 4082 *Windsor Castle* passes Southall on 15 February 1952 with the funeral train from Paddington to Windsor. *REV 63/A/2/4 / Transport Treasury*

Opposite bottom: *Courtesy Great Western Trust collection, Didcot*

A Terrier at Swindon

Two ex-LB&SCR Terriers that had previously worked on the Weston, Clevedon & Portishead Railway (WC&PR) were purchased by the GWR following the abrupt closure of the line in 1940. These were the only locomotives considered to be serviceable and both were put to work at first on the Bristol harbour lines and later at Portishead power station and at the USA army depot at Wapley Common near Yate.

No 6, formerly No 53 *Ashstead* latterly No 4 on the WC&PR, only lasted until 1948 but No 5, formerly A1 class No 43 *Gypsy Hill* later rebuilt to AIX specifications in 1919, worked until January 1950 at Taunton and later Newton Abbot where it acted as shed pilot after which it was put into store at Swindon until withdrawal in 1954. No 5 was painted in GWR livery and renumbered 5, its original number having been No 2 on the WC&PR.

Opposite top: In 1952 this stored Terrier could be found ensconced, or should that be 'kennelled', within the stock shed at Swindon with protective sacking over its chimney. *NS201092 / Transport Treasury*

Opposite bottom: In this 1949 view, also taken at Swindon, the Terrier is parked up with seemingly a plentiful supply of fire irons! *GE186*

Above: The second Terrier taken into GWR stock was former LB&SCR No 2653, latterly WC&PR No 4, seen here in a previous life crossing the road in Clevedon on 14 June 1940. *NS 208810 / Transport Treasury*

'Flying Brick' on trial

These 4 wheeled railbuses were produced by Associated Commercial Vehicles/British United Traction, ACV being the umbrella company that owned Park Royal Vehicles along with AEC, Maudslay & Crossley marques.

British United Traction was a joint venture between Leyland and AEC set up after WW2 to produce road vehicles. With the imminent modernisation of the railways they also established a rail section eventually supplying some 75% of all DMU engines and control equipment in the UK. An order was placed in June 1950 and with construction complete by the end of 1951 the unit was delivered at the end of April 1952. Nicknamed rather unkindly the 'flying brick', no doubt due to its lack of aerodynamic styling, a 3 car set was built as a demonstration train, formed of two power cars and a centre trailer, although it could also be operated as a one or two car train. The body was built by Park Royal and the underframe and mechanics by AEC who supplied its two 125hp diesel engines as used in LT buses. Initially they were numbered Cars 1-3 but were later given the BR numbers M79740-2. They were trialled in many places around the country leading to them being bought by the LMR in January 1955 with a further set and spare power/trailer car delivered in 1955 followed by another 3-car in 1957. The later eight cars (M79743-50) did not have bodyside skirts and sported sliding lights rather than droplights on the sides.

It ran its initial trials between Didcot and Newbury during w/c 28 April 1952 followed by a demonstration run at Gerrards Cross on 23 May 1952. It was then based at Neasden shed from where it worked a series of trials from Marylebone on outer London suburban routes such as the High Wycombe/Princes Risborough line. In June the cars worked on Mondays to Fridays as part of the shuttle service over the then steam worked line from Epping to Ongar. A two week stint then followed in July working over the St Albans - Watford Junction line with 1, 2 or 3 cars being operated as required. The LTE Metropolitan Line shuttle from Chalfont & Latimer to Chesham was also worked by the set in October and in the following year they undertook trials for a short period on the Southern Region's Allhallows-on-Sea branch in Kent after which they moved to the LMR where they operated the Harrow – Belmont and Watford – St. Albans branches.

Finished in grey livery Car N°1 consisted of a driving compartment and two saloons, each seating 16, together with a Guard's compartment and a luggage compartment containing a second driving position. Car N°2 had two saloons, with 24 and 21 seats respectively, and a driving compartment. Sandwiched between these two units was Car N°3 which had 52 seats spread over two saloons. Gangways were provided between the cars enabling the Guard/Conductor to check tickets and collect fares. The seats were bus-type and upholstered in a rather fetching shade of strawberry pink. Acceleration was brisk but the rather low top speed of 50mph was not really suitable for main line running. It was to be another three years before production DMUs came on stream with the ACV units being placed in store from February 1959 but seeing occasional use thereafter standing in for failed DMUs. They spent a considerable time in store at Derby before being scrapped in 1963.

Opposite top: The photographer catches the eye of the crew of the experimental 3 car DMU, with unit No 2 leading, in an image which Illustrates well the rather box like structure of this ACV vehicle. *C140AS*

Opposite bottom: As a double decker bus makes a tight turn on the road above Gravesend Central station this shot, taken on 24 October 1953, captures the arrival of the ACV vehicle which was undergoing trials on the Southern region. *CR 4847*

Right; On the same date the unit has reached Cliffe station on the line to Grain and Allhallows where it passes a steam hauled service coming in the opposite direction. A couple travelling in the DMU take advantage of the rear facing windows – not quite the "Devon Belle" Observation Car perhaps but a novelty nonetheless. The DMU will branch off to Allhallows-on-Sea at Stoke Junction halt some 7¼ miles distant. *RCR4848 / Transport Treasury*

Beeching's precursor : The Branch Line Committee

The Branch Line Committee was actually comprised of several committees headed by a national committee of Railway Executive members with one committee for each region with the remit of undertaking 'An investigation of every branch line whose earning capacity was in question'. It should be noted that individual station closures were not up for consideration at this stage. They also reviewed possible savings that could be realised with the introduction of diesel railcars.

The committees met during the period 1949 - 53 during which time approximately 200 cases were examined leading to the recommendation that 500 route miles should be closed. The Committee's title was later changed to the Unremunerative Railway Committee, widening its scope beyond consideration of only branch lines.

In 1953 the committees were disbanded after which the pursuit of closures was left up to individual regions with results varying from considerable numbers being identified in the WR but very few in the MR for example. The following table lists those branch lines and sections of line closed

between Nationalisation in 1948 and 1952, the increase in numbers being particularly evident in 1951 when the fate of some 60 lines was sealed. By the end of 1951, 142 branch lines had been subjected to analysis. The committee adopted the positive position laid down in their terms of reference such that 'Branch line policy should not be approached solely from the negative points of view of reducing expenditure. The main object was to increase or maintain net revenue and this could not be considered without regard to wider aspects now opened up by transport integration'. To this end it only recommended withdrawal of passenger services where it could offer replacement buses through its sister Road Transport Executive. It also considered the benefits of introducing lightweight diesel railcars together with the use of cheap fares to stimulate traffic growth.

From a peak rail mileage of 23,440 at the time of WW1 some 1,264 route miles had been closed by WW2 and between 1948 and 1962 a further 3,300 route miles were shut. Following the Beeching report in 1963, 4,500 route miles and 2,500 stations were closed in subsequent years.

LINE CLOSURES BETWEEN 1948 – 1952

1948

Woodford & Hinton (North Junc.) – Byfield (Woodford West Junc.)	31/ 5/48
Liverpool Lime Street – Bootle - Alexandra Dock	31/ 5/48
Rotherham (Masborough South Junc.) – Masborough Station (South Junc.)	30/ 6/48
Roxburgh Junction – Jedburgh	13/ 8/48
St. Boswells (Ravenswood Junc.) - Duns	13/ 8/48
Askern branch Knottingley – Shaftholme Junc.	27/ 9/48@
Shepherdswell (EKR) – Wingham (Canterbury Road)	30/10/48
Alne – Easingwold	29/11/48

1949

Duffield – Wirksworth	1/ 5/49 *
Lockwood (Meltham Branch Junc.) – Meltham	21/ 5/49
Broom (North Junc.) – Stratford upon Avon (SMJR)	23 /5/49
~Stonehouse – Stroud (Cheapside)	8/ 6/49 #

Dudbridge Junc. - Nailsworth	8/ 6/49 #
Gas Factory Junc. – Bow Junc.	7/11/49
Longniddry Junc. – Haddington	5/12/49

1950

Kintore Junc. – Alford	2 /1/50
Stranraer Town – Portpatrick	6/ 2/50
Mold (Tryddyn Junc.) – Brymbo	27 /3/50
Bedlington – Morpeth	3/ 4/50
Holehouse – Ochiltree (Belston Junc.)	3/ 5/50
Pickering (Mill Lane Junc.) – Seamer (Inner Junction)	5/ 6/50
Malton East Junc. – Driffield (West Junc.)	5/ 6/50
Fallside (Bothwell Junc.) – Bothwell CR	5/ 6/50
Symington Junc. – Peebles	6 /6/50
Kinross (Mawcarse Junction) – Ladybank (South Junc.)	6/ 6/50

Auchinleck – Cronberry	3/ 7/50	Marsh Lane Junc. – Aintree Station Junc.	2/ 4/51
		Maryhill – Kilsyth	2/ 4/51
Swansea St. Thomas – Brynamman	25/ 9/50	Craigleith – Barnton	7/ 5/51
Pilmoor – Knaresborough	25/ 9/50	Kelvedon – Tollesbury	7/ 5/51
Newton Stewart – Whithorn	25/ 9/50	Lampeter – Aberayron	7/ 5/51+
Queenborough – Leysdown	4/12/50	Chipping Norton – King's Sutton	4/ 6/51

1951

Llynclys Junc. – Llangynog	15/ 1/51	Grain – Port Victoria	11/ 6/51
Garforth Junc. – Castleford (East Junc.)	22/ 1/51	Hertford North – Hatfield	18/ 6/51
Kington – New Radnor/Titley Junction – Presteign	5/ 2/51$	Essendine – Bourne	18/ 6/51
Cruckmeole Junction – Minsterley	5/ 2/51	Rainford Junction – St. Helens	18/ 6/51
Beaufort – Ebbw Vale HL	5/ 2/51	St. Helens – Widnes & Ditton Junction	18/ 6/51
Newburgh – St. Fort	10/ 2/51	Towcester – Cockley Brake Junction	2/ 7/51
Quakers Yard HL – Joint Line Junc. (Merthyr)	12/ 2/51	Alyth Junction – Alyth	2/ 7/51
Whiteinch West Junc. – Whiteinch (Victoria Park)	2/ 4/51	Felixstowe Beach – Felixstowe Pier	2/ 7/51
Harrogate (Ripley Junc.) – Pateley Bridge	2/ 4/51	Kirkintilloch – Kilsyth	4/ 8/51

Rickmansworth's Church Street station sees a Watford electric set comprising Oerlikon cars at the platform. Although an intensive electric service was operated, trains were withdrawn in March 1952. *OTA 3595*

Brechin, seen here in 1966, also suffered the withdrawal of its passenger services to both Bridge of Dun and Montrose and to Forfar in 1952 saving an estimated £11,000 p.a. The station is currently occupied by the Caledonian Railway preservation society who operate trains on the four miles to Bridge of Dun. *RP1705*

Little Somerford – Malmesbury	10/ 9/51
Plymouth Friary – Turnchapel	10/ 9/51
Annbank - Cronberry via Drongan	10/ 9/51
Eskbank – Polton	10/ 9/51
Hawthornden – Penicuik	10/ 9/51
Auchengray – Wilsontown	10/ 9/51
Reston – Duns	10/ 9/51
Coatbridge – Bothwell Junc.	10/ 9/51
Selkirk – Galashiels (Selkirk Junction)	10/ 9/51
Saltburn West – Brotton Junc.	10 /9/51
Wakefield Kirkgate - Edlington	10/ 9/51
Hatfield – St. Albans	1/10/51
Balquhidder – Comrie	1/10/51
Perth (Almond Valley Junction) – Crieff	1/10/51
Blackwood Junction – Tillietudlem/Brocketsbrae – Hamilton	1/10/51
Kirkintilloch – Aberfoyle	1/10/51
Montrose – Inverbervie	1/10/51
Muir of Ord – Fortrose	1/10/51
Macduff – Inveramsay	1/10/51
Ayr – Annbank	1/10/51
Cronberry – Muirkirk	1/10/51
Sutton South Junc. – Sutton in Ashfield	1/10/51"
Oxford Rewley Road – Oxford North Junc.	1/10/51
Glastonbury & Street – Wells Priory Road	29/10/51
Highbridge – Burnham-on-Sea	29/10/51^
Chathill – Seahouses	29/10/51
Durham – Waterhouses	29/10/51
Batley West Junc. – Tingley West Junc.	29/10/51

Tingley East Junc. – Beeston South Junc.	29/10/51
Louth – Bardney	5/11/51
Llantrisant – Cowbridge	26/11/51
Chesterfield Market Place – Shirebrook North	3/12/51
Bangor – Bethesda	3/12/51
Sandling Junc. – Hythe	3/12/51
Uffington – Faringdon	31/12/51
Much Wenlock – Craven Arms	31/12/51

1952

Aintree Central – Southport Lord Street	7/ 1/52
Lowton Junction – St. Helens Central	3/ 3/52
Bishop's Stortford Junction – Braintree & Bocking	3/ 3/52
Croxley Green Junction – Rickmansworth Church Street	3/ 3/52
Coryton – Corringham	3/ 3/52
Weymouth Junction – Easton	3/ 3/52

Coxhoe Junction – Spennymoor	31/ 3/52
Norton South Junction – Ferryhill South Junction	31/ 3/52
Mywyndy Junction Llantrisant – Tonteg Junction	31/ 3/52
London Road Junction – Leith Central	7/ 4/52
Stratford-upon-Avon (S&MJR) – Blisworth	7/ 4/52
Tottington Junction – Holcombe Brook	5/ 5/52
Lockerbie North Junction – Dumfries No 1 Junction	19/5/52
Heacham Junction – Wells Junction	2/ 6/52
Hadley Junction – Coalport East	2/ 6/52
Cemetery North Junction Hartlepool – Ferryhill North Junction	9/ 6/52
Castle Eden Junction – Murton Junction	9/ 6/52
Amesbury Junction – Bulford : Grateley Junction-Newton Tony	30/ 6/52
Launceston North Junction spur – Launceston North	30/ 6/52

Bulford station sees U class mogul No 31611 with a freight service awaiting departure for Amesbury on 23 October 1959. The passenger service, such as it was – latterly just one train each way on weekdays - ceased in June 1952. *JH1152*

Haughley – Laxfield	30/ 6/52	Holmes Junction – Rotherham Westgate	6/10/52
Clydach Court Junction – Old Ynysbwl Halt	30/ 6/52	Wickham Market Junction – Framlingham	3/11/52
Montrose North Junction – Dubton North Junction : Brechin – Bridge of Dun South Junction : Brechin South Junction – Forfar South Junction	4/ 8/52	Kirkby Stephen Junction – Tebay	1/12/52
Kirriemuir Junction Forfar – Kirriemuir	4/ 8/52	Malvern & Tewkesbury Junction – Upton on Severn	1/12/52
Bothwell – Hamilton	15/9/52	Edington Junction – Bridgwater North	1/12/52
Reedsmouth Junction – Morpeth : Scotsgap Junction – Rothbury	15/ 9/52	Upwey Junction – Abbotsbury	1/12/52
Wroxham Junction – County School	15/ 9/52		
Leominster – Bromyard	15/ 9/52		
Pencader Junction – Newcastle Emlyn	15/ 9/52		
Elsenham East Junction – Thaxted	15/ 9/52		
Belmont – Stanmore Village	15/ 9/52		
Merstone Junction – Ventnor West	15/ 9/52		

*	Services suspended 14/7/47 and never reinstated
#	Services suspended 14/7/47 and never reinstated
+	Last train 12/2/51 with official closure 7/5/51
^	Used for advertised excursion trains until 8/9/62
$	Services temporarily withdrawn but never reinstated
"	Closed 1/1/17, Re-opened 9/7/23, Closed 4/5/26, Re-opened 20/9/26. Workmen's service, operated until 1/10/51
~	Used for diversions until 3/1/66
@	Open for freight and diversions. Passenger service re-introduced 2010

The branch line between Upwey Junction and Abbotsbury was one of those recommended for closure by the Branch Line Committee and duly saw the end of passenger services in December 1952. This view of the intermediate station at Upwey, which continued to see freight traffic until 1962, was taken shortly before the remaining track was removed in 1965. *RP1263*

The first stage in the wholesale contraction of the rail network on the Isle of Wight occurred with the closure of the Merstone - Ventnor West line in September this year.

The SR announced that it intended to close the Newport – Freshwater, Newport - Sandown and Brading – Bembridge routes within the next year followed by the Cowes – Ryde line within the next five years. In fact the latter route lasted until April 1966 although in that month there was to be another closure involving the section of line between Shanklin and Ventnor leaving just the 8½ mile rump between Ryde Pier Head and Shanklin that we have today.

Propped up against the bicycle racks at London Bridge station was this notice advertising the forthcoming closure of the Ventnor West branch. One might question the relevance of this closure to passengers at London Bridge; if they had but only known it was a precursor of what was to come. *LSDC1439*

A few passengers await the arrival of the next service in this view of Ventnor West dating from the early 1950s. The authorities have kindly added the information that the station is "168 feet above sea level" – not something to draw attention to perhaps when you are flogging up the hill to the station from the beach! It was also situated some distance to the west of the town centre, the original plans to bring it nearer being thwarted by newly built housing and the railway company's parlous financial state. *LS64665*

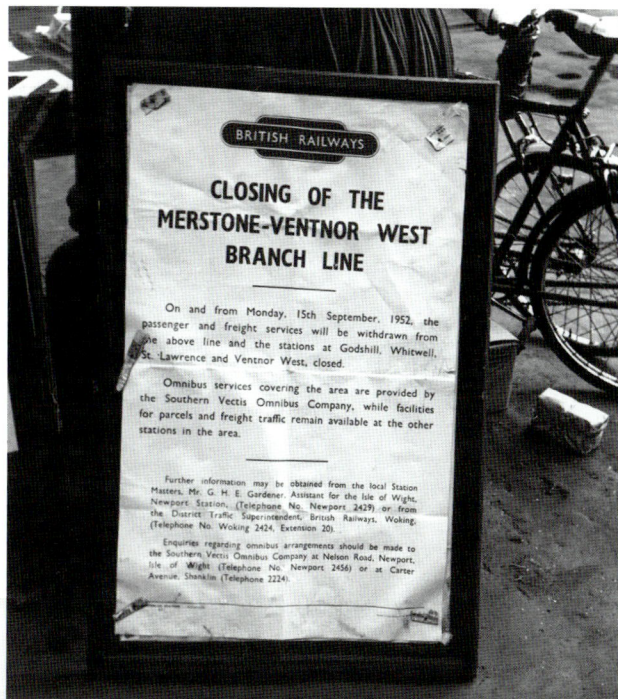

BRITISH RAILWAYS

CLOSING OF THE MERSTONE-VENTNOR WEST BRANCH LINE

On and from Monday, 15th September, 1952, the passenger and freight services will be withdrawn from the above line and the stations at Godshill, Whitwell, St. Lawrence and Ventnor West, closed.

Omnibus services covering the area are provided by the Southern Vectis Omnibus Company, while facilities for parcels and freight traffic remain available at the other stations in the area.

Further information may be obtained from the local Station Masters, Mr. G. H. E. Gardener, Assistant for the Isle of Wight, Newport Station, (Telephone No. Newport 2429) or from the District Traffic Superintendent, British Railways, Woking, (Telephone No. Woking 2424, Extension 20).

Enquiries regarding omnibus arrangements should be made to the Southern Vectis Omnibus Company at Nelson Road, Newport, Isle of Wight (Telephone No. Newport 2456) or at Carter Avenue, Shanklin (Telephone 2224).

Opposite top: Taken from a similar viewpoint, this is a view of the RCTS 'Isle of Wight' special hauled by Class O2 W32 *Bonchurch* on 18 May 1952. W3 *Ryde* and W14 *Fishbourne* also handled other sections of the tour which covered all IOW routes except that from Sandown to Ventnor. *LS64654*

Opposite bottom: Back to April 1949 for this view of Terrier W13 *Carisbrooke* in malachite livery with "British Railways" on the tank sides with a single pull push coach operating the branch service to Merstone, the last line to open on the island. *LS PMA1288*

Above left: An interesting three-arm home signal surviving at the entrance to Ventnor West station. Where stop signals were placed below each other on the same post they applied to lines with the following rule, 'top to bottom is left to right'. *LS PMA1291*

Above right: Another ancient artefact was this ground signal at Ventnor West photographed on 18 April 1949. *LS PMA1384*

Right: No surprises for expecting that in later years there would probably be no Sunday service on this bucolic branch line although this fact apparently warranted the erection of this helpful sign at the intermediate station of Godshill. Three Sunday services each way were provided in pre-war summers but rationalisation had seen their withdrawal as an economy measure. *LS PMA1364*

The first station out of Ventnor West, and indeed the temporary terminus from 1897 until extension to Ventnor followed in 1900, was St. Lawrence, perched awkwardly on the hillside. *LS PMA1361*

As an economy measure the former passing loop and signal box were removed at Whitwell in 1928. A couple of hikers and a pair of milk churns constitute the sole occupants of the platform which also houses scaffolding put up apparently to effect some roof repairs. The isolated nature of the station is apparent in this view. *LS64683*

Calling at the attractive station of Godshill on 16 April 1949 is Class A1X No W8 *Freshwater,* carrying duty number 22, with its single coach forming a service from Ventnor West to Merstone. *LS PMA1365*

Merstone was where the Ventnor West branch reached civilisation of a sort as passengers could change here for Newport and Sandown and the rest of the island network. W36 *Carisbrooke*, seen at the branch platform, awaits departure time for Ventnor West. The signal box controlling the adjacent level crossing contained 28 levers with two spare and the station, apart from being a junction, despatched a considerable quantity of agricultural produce from the nearby fertile Arreton valley with, at one time, three trains of sugarbeet being despatched daily to Medina Wharf for onward transit to Selby in Yorkshire for processing. Another unusual feature of the station was the provision of a subway upon the station's rebuilding consequent upon the opening of the line to Ventnor West. *LS64796*

Continuing our series of the more unusual and bizarre happenings on BR during this year

SIR TOPHAM-HATT. The headgear of the 'Fat Controller' in the Rev. W Awdry's Thomas the Tank Engine series no doubt led to his name although perhaps in these hypersensitive days he would not be 'fat shamed' by such an adjective. In 1952 rumours were abroad that BR would shortly be ceasing the provision of silk hats to stationmasters at principal locations. Rather, some revisions to this somewhat

archaic tradition have been proposed in that new silk hats will not automatically be provided to new incumbents once their predecessors have retired. Some additions to the principal stations list, which now numbers 26 in total, 10 of which are located in London, have been made in that Charing Cross, Birmingham Snow Hill, Bristol Temple Meads and Cardiff Central will now enjoy this privilege. The order remains that silk hats and morning dress should be worn by stationmasters on suitable occasions such as visits by royalty or other VIPs.

Sporting his 'topper', the Liverpool Lime Street stationmaster chats to the crew of D211 heading the 'Red Rose' express in 1959. *Maurice Hudson*

EFFING (HAM) & BLINDING. The crew of Lord Nelson Class No 30854 *Howard of Effingham* no doubt let fly with a few choice expletives as their mount derailed south of Shawford on 20 July whilst working the 3:24pm stopping service from Southampton to Waterloo. At the convergence of a four track into a two track section the crew mistook the signals relating to the up main line for those relating to the up relief line, on which the 4-6-0 was travelling at the time. Realising their mistake too late the locomotive ran into a sand drag at 20mph, plunged down the embankment

and turned over onto its side. Fortunately no passengers or crew were injured and although the tender was retrieved within a short time, initial attempts to lift the locomotive proved unsuccessful until, using jacks, it was tipped into a pit prepared alongside. Using Kelbus apparatus, in a manner similar to that employed in the Cocking derailment of the previous year (See article in *Railway Times No 4*) and a winch on a bulldozer, the recalcitrant 'Nelson' was pulled up the slope previously prepared some 10 days after the initial mishap.

Dramatic scenes following the toppling of No 30854 near Shawford. *S C Townroe*

FOR WHOM THE BELLE TOLLS. An announcement in 1952 that the 'Devon Belle', introduced in June 1947, would not be reinstated that year was followed a week later by a further bulletin that in fact it would continue to run but only on summer weekends which caught out the Exeter 'Express & Echo' which carried an obituary in its pages for the service in March 1952 which proved in the event to be somewhat premature. The economics of running the all Pullman 'Devon Belle', which often consisted of 12 heavyweight Pullmans plus the Observation Car, were however under severe scrutiny. For the first year of operation the train ran on Fridays, Saturdays, Sundays and Mondays until October 27 with separate portions for Plymouth and for Ilfracombe, the latter including the Observation Car. The 1948 season ran from 14 May – 26 October with the addition of a down train on Thursdays and a balancing up train on Tuesdays leaving Wednesday as the only day without a service. Although

reports often spoke of '14 well-filled Pullman coaches' in the first year or two of operation, in truth the Belle's initial impact began to fade as early as 1949 when that summer's revenue failed to meet operating costs.

As patronage had never been as great on the Plymouth route it was decided to terminate this portion at Exeter for the 1950 season. Thus 1952's operations were limited to a down train on Fridays, both up and down workings on Saturdays and Sundays and an up train on Mondays. A couple of years later in 1954 the load, even in August, would be no more than 5 cars plus the Observation saloon. This was to remain the pattern until what were to prove to be the final runs, down on Saturday 15 September that year and up the following day. Although no public announcement was made, the death knell for the 'Devon Belle' had sounded and the train did not appear in the 1955 schedules.

The 'Devon Belle' tackles the 1 in 36 of Mortehoe bank with the up service headed by No 21C101 *Exeter* in the first year of operation, 1947. *David P Williams colour archive*

BATTERSEA FLATS. London County Council is considering naming several blocks of flats in Battersea after prominent locomotive engineers of the former constituent companies of the Southern Railway who had associations with this district of the capital. (Ed: The Patmore Estate's 28 blocks of flats subsequently built in the 1950s do indeed bear the names of locomotive engineers although not all have local associations. However, the LC&DR, SER, LB&SCR and L&SWR are represented by some of the blocks named after - Hookham, Kirtley, Martley, Mills, Banister, Billington (in error for Billinton), Mansell, no not another error for Maunsell but rather the carriage superintendent and later Works Manager at Ashford for the SER, Marsh, Morgan, Statham, Cudworth, Crampton and Beattie.) The Patmore Estate, as the development is called, is a garden estate of some 860 homes, run by the Patmore Cooperative, and is situated within the Vauxhall-Battersea-Nine Elms opportunity area in close proximity to the new American embassy, Battersea Power Station and Covent Garden Market. The estate comprises 28 red-brick apartment buildings between three and six storeys in height arranged around well-designed and maintained courtyards containing children's playgrounds and landscaped communal gardens. The buildings cover 11 accommodation types ranging from terraces of maisonettes to larger L shaped blocks and are distinguished by their innovative use of materials and by the provision of balconies.

A 'CAFF' ON RAILS. A new concept recently introduced is the 'Cafeteria Car', the prototype of which was converted at Eastleigh Works from a former LNER open 3rd. class saloon which has recently been placed in service in the north of England. It is intended to supplement, not replace, the more traditional restaurant and buffet cars and cater for the trend towards serve yourself meals. It is anticipated that these cars will be mainly used on excursion trains where a large number of customers require serving in a limited time. Collecting their food and drink from a service counter they can either take it to tables in the cafeteria car or return to their compartment seats. As previously mentioned these cafeteria conversions were designed and undertaken at Eastleigh and involved a number of former 3rd. class sleeping cars and restaurant/kitchen firsts. They were allocated to all four regions.

Cafeteria car No SC673E allocated to the Scottish Region but seen here at Clapham Junction.

'LIGHTS, CAMERA, ACTION' AT TITFIELD . Perhaps one of the most famous films of all time featuring railways was 'The Titfield Thunderbolt' whose storyline needs no repetition here. Ealing Studios took over part of the former Camerton branch line for filming during 1952 using *Lion* the ex-Liverpool & Manchester 0-4-2 locomotive as the star attraction. What is less well known is the scale of the search for a suitable branch line on which the film crew could descend. Apparently surveys were undertaken on the suitability of the following bucolic lines – Kelvedon & Tollesbury Light Railway, East Kent Railway, Kent & East Sussex, Maidenhead – High Wycombe and the Lambourn branch, not to mention various branch lines in Suffolk. However the scenic beauty of the Limpley Stoke – Camerton line won out for not only was the countryside glorious hereabouts but the villages featured the photogenically attractive Bath stone in many of their buildings. A survey by the WR civil engineers revealed no serious problems with the remaining trackwork which had played host to the occasional freight train as recently as the end of 1951.

Star of the subsequent film, *Lion* is seen at Llandudno in 1938 during the time of its appearance at the centenary of the London & Birmingham Railway which occurred during that year.

GOING TO THE FLICKS. Continuing the cinema theme the BTC Films Section has produced a series of instructional films for permanent way staff explaining the advances in mechanisation which will be shown to staff throughout the country. For those working in remote districts a twelve-month tour has been arranged bringing the films to them via the medium of travelling cinema coaches which can house up to 60 staff at a time usually under the supervision of an instructor who can respond to questions and promote discussion. Some of these modified coaches and attendant generator vans lasted in service into the 1970s.

Opposite top: Cinema coach No DS1308 and Van DS 1309 are parked in Shields Bank sidings in Glasgow in this view dated to 19 March 1958. *WS3005*

Opposite bottom: DM395017M 'Cinema coach No 2', formerly a west-coast sleeper, is seen at Wood Street siding on 12 November 1960. *F5295*

GETTING THE WIND UP AT DAWLISH. The vulnerability of the railway in the Dawlish area is well known especially following the breaching of the line in 2013. Back in 1952 concerns were raised about the resilience of the line at Dawlish Warren where a line of sand dunes protected the former GWR route from Exeter to Newton Abbot from the ravages of the sea. The Soil Mechanics Laboratory, one of the lesser known departments of the WR's Civil Engineer's Office, has installed an anemometer at Dawlish Warren to record wind speeds and directions which will be used to study the effects of wind erosion on the nearby dunes which in conjunction with wind tunnel tests will be used to establish what, if any, remedial measures will need to be taken on site to protect the line. The installation consists of a 35 ft. high pylon linked by cable to an adjacent hut containing the electrical recording apparatus. The 1947 Transport Act had required that the BTC should conduct such scientific research as was appropriate, having obtained the approval of the Ministry of Transport for the direction and type of research that was proposed.

A committee was set up under the chairmanship of Sir William Stanier, then aged 71 and retired, to enquire into the nature and extent of the research then in hand within all the Executives of the BTC and to make recommendations for the organisation of research in the future. Back in 1942 a Soil Mechanics Section had been established whilst the GWR wartime headquarters were at Aldermaston. After the war the Section moved to Westbourne Terrace near Paddington Station and in 1953 to the basement of 66 Porchester Road near Royal Oak Station. At this time the Section was much involved with the introduction of long welded rails and their expertise was used to investigate a number of factors pertinent to their introduction.

CHANGING HORSES. Some 8,000 mechanical horses are currently in operation on BR but their supremacy may be challenged by the recent experimental introduction of electric battery horses. These vehicles which have a range of 25 miles and a maximum speed of 18mph on the flat are rated as 2 tonners and haul a standard mechanical horse trailer. 100 of these 4 wheeled electric vehicles are being tried out in 16 cities and towns throughout the country in a trial that would surely meet with the approval of the environmental lobby today.

An early BR mechanical horse numbered GG2301 seen in March 1948. *R E Vincent 32/B/6/3 / Transport Treasury*

'Electric horse' in Southern Region service. Advertised on the trailer is the early 20thC tonic 'Phosferine' apparently a proven remedy for Indigestion, Maternity Weakness, Lassitude, Sciatica, Neuralgia, Loss of Appetite, Exhaustion, Mental Exhaustion, Anæmia, Hysteria, Backache, Rheumatism, Sleeplessness, Neuritis, Headache, Influenza, Nervous Debility – in fact almost any ailment!

ANOTHER WEST COUNTRY NAMEPLATE LENGTHENED . Following the report in Railway Times No 3 regarding the upgrading of West Country Pacific No 34092 from *Wells* to *City of Wells*, No 34107 *Blandford* has now been renamed *Blandford Forum* following the casting of new plates at Eastleigh Works and fixing to the Pacific on 24 October 1952

after a campaign led by an S&D railwayman from Blandford station reflecting the former name of the borough. The following year the station also benefitted from a name change, becoming Blandford Forum in September 1953 until closure in 1966.

Running in to Basingstoke station with an up cross country service to Birkenhead is No 34107 still sporting its *Blandford* nameplate in this 1951 image. *LSDC0988*

No 34107 sporting its revised nameplate.

WARTHILL EXPERIMENTAL CROSSING. Six miles from York, on the route to Hull, a new type of experimental lifting barrier has been installed at the level crossing adjacent to the station in the hope that these types of barriers will reduce the heavy maintenance costs associated with mechanically operated crossing gates. Each barrier consists of a tapering steel tube, suspended from which is a curtain of light alloy rods. The barriers are operated via a rack and pinion mechanism in the signal box. Stop boards, illuminated at night, are rotated through 90 degrees to warn oncoming traffic and mounted on the posts carrying the boards are two electric lights, one of which emits a steady red aspect when the barriers are down. The area of the crossing is also floodlit at night.

The lifting barriers are raised in this view of the level crossing at Warthill. *NS 201904 / Transport Treasury*

Centenary Trio

1952 marked the centenary of two important events in the history of the GNR and of the opening of Brighton Works

In 1852 the GNR's main line was completed with the opening, in August, of the 'Towns Line' between Peterborough to Retford via Grantham and Newark and, in October, of their new London terminus at King's Cross. A centenary exhibition was held in Grantham at the Guildhall, opening on 28 July 1952 whilst the 'Towns Line Centenary Exhibition' opened at Retford in September 1952.

The exhibition was opened on Wednesday 24 September by Sir Ronald Matthews, the last chairman of the LNER who travelled by train from King's Cross to Retford hauled appropriately by the locomotive which bore his name, No 60001 stopping en route to pick up civic representatives of 'the Towns', at Peterborough, Grantham and Newark.

Another centenary marked this year was that of Brighton Works which had opened in 1852. To mark the occasion a special all Pullman train from London to Brighton, to a 60 minute timing, was organised by the RCTS for 5 October headed by Brighton Atlantic No 32424 *Beachy Head*. Some 300 tour participants visited the shed and works and Terrier No 32636, the oldest locomotive on the Southern Region, operated a shuttle to Kemp Town. The tour was repeated on 19 October. All this for 22/6d (£1.22½p)!

The 'Centenaries Express', organised by amongst others a certain A F Pegler, ran from King's Cross to York and return on 28 September headed by No 60007 *Sir Nigel Gresley* with a scheduled time of 3 hours 15 minutes which was the fastest time attempted since 1939 which in fact it bettered by 1½ minutes, achieving a top speed of 87½ mph travelling up grade at Little Bytham.

The Centenaries Exhibition was held at King's Cross from 13 to 18 October, the station having opened in 1852. The frontage of the station was floodlit and two preserved GNR locomotives, No 1 and No 251, were on show having been brought to London from the railway museum in York. World steam record holder No 60022 *Mallard* was also on view and completing the illustrious line-up were a GNR dining car and a modern BR kitchen car.

Awaiting departure time at King's Cross is A4 Pacific No 60007 *Sir Nigel Gresley* adorned with a commemorative headboard. *OTA 20994*

Brighton Works also celebrated its centenary in 1952 and this view taken from a favourite location of railway photographers, Howard Place on the chalk hillside, looks out towards the substantial Works buildings. An unidentified mogul is seen backing out of the station whilst a Schools class 4-4-0 is parked on one of the shed roads. *LSDC 3766*

RAILWAY TIMES 1952

Top: What the photographer describes as the RCTS 'pack', all of course wearing jackets and ties, wandering through the still active Brighton Works having been conveyed by the special train of 5 October 1952. *JCF L1-3*

Middle: Returning the RCTS members on the eight vehicle Pullman special to London Victoria is Brighton Atlantic No 32424 *Beachy Head*, now of course stunningly resurrected as a new build on the Bluebell Railway. The train had taken the route London Victoria - Battersea Park - Pouparts Jn - Clapham Junction - Balham - Streatham Common - Norbury - Selhurst - Windmill Bridge Jn - East Croydon - Purley - Coulsdon North - (via Quarry line) - Three Bridges - Brighton. Notes on the 'Six Bells Junction' site report, *'Outward journey to Brighton was completed in 58m 48s, the return in 60m 13s for the 50.9 miles. There was a permanent way check in the East Croydon area in the down direction which caused a loss of time between Windmill Bridge and Coulsdon North but this was later made up. The up train reached 80 mph near Horley but was checked in the London area, but nevertheless arrived on time.'* LSDC1502

Bottom: Whilst in Brighton the tour participants were offered the opportunity of a trip out to Kemp Town where Class A1X No 32636 hauling pull push set No 727 is seen about to enter the long closed terminus with one of the two shuttles operated that day. *LSDC 1497*

Crabs & Winkles - what about the Oysters?

Whitstable is famous today for oysters which have been harvested locally since Roman times, but in 1952 the branch from Canterbury to the town, known as the 'Crab & Winkle' line, reflecting the initials of the company (C&W) and as a nod to the town's shellfish industry, saw the withdrawal of its remaining freight service.

Dating from the very early days of railways the Canterbury and Whitstable Railway was opened as long ago as 1830, being one of the earliest lines in the country. However, the line was never prosperous, even under later SER management, and once the LC&DR opened its line in 1860 there was a much improved passenger service from the town

to London. It came as no surprise that passenger services were withdrawn as early as 1931 and the final scheduled freight train ran on 29 November 1952 although there was a short reprieve during the floods of February 1953 when the line was reopened from 5 February to 1 March in order for traffic to bypass the main line between Whitstable and Faversham damaged in the flood.

Track lifting followed soon afterwards. The limited clearance of Tyler Hill tunnel necessitated the cutting down of the chimneys and domes of several members of the Stirling R1 class of tank locomotive which became synonymous with the line.

Top: Class R1 No 31339 with cut down chimney and dome shunts wagons adjacent to the old station at Whitstable. *LSDC1528*

Bottom: In this view taken from the level crossing near Whitstable Harbour station looking towards Canterbury on a very inclement day, the main station building appears to have been demolished. The Sea Scouts and a Telegraph Office have taken over an adjacent former railway building. *LSDC1534*

Top: View of Whitstable Harbour with sheeted bulk grain wagons visible on the quayside.*LSDC1532*

Bottom: This sheeted grain wagon has a Morton brake which was a popular design which became part of the Railway Clearing House (RCH) standard specification for private owner rolling stock of 1923. To save money the hand brake was often only fitted on one side of the wagon, the other side having wheels with no brakes at all. The wagon is lettered 'When empty return to Whitstable Harbour' whilst the sheeting carries the message 'Grain Traffic Only'. *LSDC1531*

In addition to the three members of the R1 class which retained their rounded Stirling cabs and had their chimneys and domes cut down, at least a couple more had these items retrofitted including No 31339 seen here leaving a rain soaked Whitstable Harbour in 1952 with an afternoon freight bound for Canterbury. Note the collection of 'dragon's teeth' concrete tank traps on the left of the picture. *LSDC1535*

The reason for the reduction in height of the chimney and dome of the R1s that operated over this route was brought about by the restricted bore of Tyler Hill tunnel. *LSDC1551*

Cross Dressing Diesels
(Drewrys with skirts!)

BR have placed contracts for 13 Drewry diesel mechanical shunters of 204 bhp, four of which have been adapted for use on the Wisbech & Upwell Tramway and on dockside lines.

The four tram locomotives intended for use on the W&U tramway were constructed by the Vulcan Foundry Co. as part of the Drewry order. They have been fitted with side skirts and cowcatchers as much of the route is unfenced. They also have a speed governor limiting top speed to 12 mph although this can be disabled when the locomotives operate over other sections of the line, allowing a heady 30mph to be attained. Two of the quartet have also been fitted with automatic vacuum brake equipment for train application. Numbered 11100-3, one has been allocated to Ipswich, another to Yarmouth whilst the remaining two will go to March depot for use on the W&U tramway. An inaugural run on the tramway was operated by No 11102 on 4 June.

Two of the skirted Drewrys, Nos. 11101 and 11102, are seen in the yard at Wisbech North on 9 June 1953. *REV75/C/2/2*

No 11103 shunts at Yarmouth Docks on 20 August 1957. *ES4087*

GWR Railcar goes a-wandering

BR were rather late in looking at the potential of diesel railcars, the concept having been in use on the continent of Europe for some years before BR took action with authority being given in September 1952 for the expenditure of £0.5m for multiple unit trains. It was the intention to trial these units in the West Riding of Yorkshire. Prior to this announcement the unusual sight of an ex-GWR railcar could be seen in the Eastern and North Eastern regions of BR. Railcar No W20W was the chosen vehicle and in the summer of 1952 was to be seen around Wakefield, Doncaster and Bradford before beginning a 4 day stint working from Boston to Grimsby, Skegness and Mablethorpe with a view to establishing running times on some of the Lincolnshire branch lines. In October it was seen at Lincoln and also visited Immingham Docks after which it returned to the WR where it continued in service for a further 10 years. It now has a home on the K&ESR, one of 3 such railcars that made it into preservation, the others being housed at the NRM (No 4) and the GWS at Didcot (No 22).

Above: Ex-GWR Railcar No W20W was captured on film at Bradford Exchange on 8 August 1952 during its sojourn in the North. *G531*

Bottom left: Another wandering railcar during 1952 was W14W which formed the 8.50am departure from Solihull to Buxton on 28 June 1952 operating a special for the Solihull Society of Arts running via Bordesley, Burton and Uttoxeter to Buxton. It is seen here passing Birmingham's Lawley Street goods yard. *1001*

Bottom right: W20W is seen here in the late 1960s at Rolvenden, having arrived on the K&ESR in 1966. It is keeping company with Stroudley A1 class formerly No 70 *Poplar*, latterly *Bodiam*, and diesel electric locomotive Ford No 1 which also arrived on the K&ESR in 1966. Interestingly the railcar, which is currently undergoing restoration, was originally one of just two, and is now is the only surviving, 1940s diesel railcar to be constructed with a dual range gearbox which facilitated operation on the slower speeds of branch lines and the faster speeds of main lines. *Editor*

Last of the I3s

The last survivor of Marsh's I3 class of 4-4-2Ts, No 32091, has been withdrawn. Entering traffic in 1913, this was the last of the class to be built and the last to undergo a general overhaul at Eastleigh in December 1949. Having spent some time at Brighton out of use during 1952 it made its last run in steam on 6 December 1952 when it ran light engine to Ashford Works where it was scrapped the following month, achieving the distinction of being the first ex-LB&SCR locomotive to be broken up at Ashford.

During their career the superheated I3s had successfully worked express services over much of the LB&SCR network, being particularly associated with the Mid Sussex line. In spite of the availability of more modern locomotives in early Southern Railway days the I3s continued to put in appearances on prestige services such as the 'Southern Belle'. However, with increasing electrification their duties became fewer on the main line and they were modified to operate over the Eastern section with its restricted loading gauge working semi fast services to the Kent coast and inter regional services such as the 'Sunny South Express'. Race day specials also saw the I3s in action and on the outbreak of war two were even loaned to the GWR at Worcester, returning to the Southern in 1943.

All members of the class bar one, No 2024 which was withdrawn in 1944 with badly cracked frames, were inherited by BR in 1948. Hauling heavy commuter trains over the Oxted, Uckfield and Tunbridge Wells lines they continued to give sterling service but in January 1950 it was decreed that they should be withdrawn from service when heavy repairs were required. With the arrival of increasing numbers of new 2-6-4Ts on the SR in 1951, the fate of the I3s was effectively sealed.

Opposite top: No 32091, which was destined to be the last survivor of the class, is seen on Brighton shed during 1952 *NS2008808B / Transport Treasury*

Opposite bottom: An image dating from the early years of the 20th century reveals No 29 looking splendid in LB&SCR umber livery at New Cross Gate.

This page: No 32090 is seen at Eastbourne shed in 1948 having received its 'British Railways' marking in July of that year. It would be withdrawn from Three Bridges depot at the end of 1950. *NS200880A / Transport Treasury*

BR	Date to service	Date withdrawn	BR	Date to service	Date withdrawn
32021	10/1907	10/1951	32079	11/1910	4/11/1950
32022	03/1908	5/5/1951	32080	12/1910	3/1950
32023	02/1909	28/7/1951	32081	12/1910	9/1951
32025	03/1909	1/1950	32082	08/1912	23/6/1951
32026	03/1909	9/1951	32083	08/1912	23/6/1951
32027	05/1909	2/1951	32084	08/1912	3/3/1951
32028	12/1909	10/1951	32085	06/1912	24/6/1950
32029	12/1909	3/1951	32086	09/1912	6/10/1951
32030	03/1910	9/1951	32087	11/1912	11/1950
32075	03/1910	12/1951	32088	11/1912	11/1950
32076	03/1910	9/12/1950	32089	12/1912	14/4/1951
32077	10/1910	3/1951	32090	03/1913	12/1950
32078	11/1910	2/1951	32091	03/1913	6/1952

No 32023 runs onto Brighton shed in 1951, being withdrawn at the end of July in that year. *NS200529A / Transport Treasury*

A prestigious duty for an I3 was the Derby Day Pullman special to Epsom Downs, seen here at Victoria in the capable hands of Southern Railway No 2087, latterly BR No 32087. There would be several such First Class specials run for this famous race meeting in pre-war days. *MTP067-00012 / Transport Treasury*

Last Post for the 'Bisley Bullet'

The National Rifle Association (NRA), which was designed to encourage rifle shooting proficiency amongst the rifle and artillery voluntary corps formed following a war scare with France in the 1850s, was for many years located on part of Wimbledon Common.

Forced to vacate this site in 1888, the LSWR were keen to keep the NRA's new headquarters within their own operating area and, by offering to help with construction costs and by making cheap fares available, they managed to ensure that a site at Bisley near their main line at Brookwood was chosen. Even though trains would generally run for only one month during the year, when the NRA's annual meeting was held in July, it was still considered a worthwhile investment and the 1¼ mile branch line opened in 1890. Services were later introduced at other times of the year including a Saturday afternoon service which ran throughout the April – October shooting season from 1900-1914.

The operation of the occasional special and troop trains was supplemented by irregular goods services and trains for shooting competitions held at Bisley as part of the Olympic Games in both 1908 and 1948. The LSWR agreed to provide the locomotive and carriages to work the services in return for 50% of the gross receipts but as it was only used for these limited periods, and in times of national crisis during wartime, it is perhaps not surprising that it was never a great money spinner.

At the start of the First World War however, some 150,000 troops underwent training at Bisley in just four months, many coming by train. Track extensions were built totalling 3 miles from Bisley Camp to serve other camps built at Pirbright, Deepcut and Blackdown. There were also three narrow gauge tramways, one carrying competitors to a distant firing range, another carrying targets to shooting butts and one for carrying targets on a short stretch of track.

There were originally two stations at Bisley Camp but in 1891 a new centrally located station was decided upon which consisted of one long brick faced platform with a passing loop. The unassuming timber station building comprised a waiting room, booking hall and office. Crossing gates were positioned at the end of the platform to accommodate one of the camp's internal roads and three sidings and an unloading dock were also provided. The platform was subsequently extended and a canopy added at right angles to the station building. 1930s economies saw some rationalisation of the track layout which required subsequent pull-push operation of trains, generally with a 2 coach set, thereafter. Trains left from a branch platform at Brookwood. The military camp extension lines did not last long following the end of hostilities and the track was lifted in 1928. However about one mile was relaid during World War II but again this did not survive the outbreak of peace very long.

Waiting at the branch platform at Brookwood with its two coach set, No 734 is an M7 tank bound for Bisley Camp during 1952, the last year of operation of the branch. This pull push set was one of nine such sets converted from ex-LSWR 46' 6" 'Emigrant' corridor stock in 1942/3. The sets comprised a Driving Brake Third and a Composite and they retained a corridor connection between the two coaches. *LSDC1402*

With expenses and maintenance costs exceeding receipts after the First World War no further payments were made to the NRA until the 1930s. With deteriorating track and the ever increasing cost of maintaining the service, the possibility of replacing trains with buses was investigated but was not pursued and the track upgraded. After proving an asset again during World War II the branch became an increasing financial liability and 1952 saw the last rail served NRA annual meeting and the 'Bisley Bullet', as the train had been affectionately dubbed, operated for the final time on 19 July of that year.

The name Bisley was to be reincarnated in a sense a few years later when the tv series 'The Army Game', which ran to 155 episodes over four series from 1957-61, starred Bill Fraser as Sgt. Major Claude Snudge and Alfie Bass as Pte. Montague 'Excused Boots' 'Bootsie' Bisley.

On 23 November 1952 one of the elderly Adams 'Jumbos' of Class 0395, No 30577, heads the 'Bisley Tramway & North West Surrey' RCTS railtour which was booked to leave Waterloo at 12.38pm with some 300 passengers on board the 7 coaches comprising three ex-L&SWR corridors and two ex-Sheppey articulated twin sets. On arrival at Brookwood the passengers were split into two groups for their trip on the Bisley branch using two coach pull push set No 263 hauled by M7 No 30027. Introduced in 1881, just one of the 18 'Jumbos' constructed had been withdrawn by the date of this image with No 30577 lasting in traffic until February 1956 having put in 72 years service. *LSDC1539*

No 30027 and its set have just passed over the Basingstoke Canal by means of this girder bridge. *LSDC1406*

With a severe speed restriction applying as indicated by the notice board and necessitating the provision of a check rail, No 30027 rounds a curve on the Bisley branch with tour participants. *LSDC1409*

Top: "Sure your flag is big enough?" might be the comment made to the squaddie doing the honours as No 30027 rumbles over this ungated crossing on the branch. *LSDC1408*

Bottom: Journey's end at Bisley camp station. The platform and station still exist but as the headquarters of the Lloyds Bank Rifle Club. A Mark 1 sleeping car has been positioned as dormitory accommodation on a length of track by the platform as a fitting reminder of the railway. *LSDC1410*

Necropolis Necrosis

Resulting from the overcrowding of London cemeteries apparent in the middle of the 19th century the London Necropolis Company (LNC) was established in 1853 when a private terminus was built outside Waterloo station and the line from Brookwood to Brookwood cemetery constructed, the cemetery being consecrated in November 1854.

Regular funeral trains were then operated using L&SWR locomotives although some of the rolling stock was owned or leased by the LNC even though it was painted in the livery of the L&SWR or Southern Railway. This stock included six hearse vans, each of which could accommodate 12 coffins split over two levels. In 1899 two more hearse vans were constructed carrying some 24 coffins split over three levels. In all cases even the coffins were class segregated as we could not have the 'hoi polloi' dead mixing with the 'toffs'. These later vans were numbered 377 and 378, being renumbered by the Southern Railway 1425 and 1426 with the last of these vans, No 1426, still bearing SR livery in existence as late as 1950.

The original London terminus was located in York Street and remained in use until 1902 when a new building and platform opened in Westminster Bridge Road following improvements to Waterloo's layout. This latter site remained in use until April 1941 when it was hit in the heaviest bombing raid endured so far in the London Blitz. The station closed for good the following month with the final funeral train, conveying the body of a Chelsea pensioner, running on 11 April.

Although some funeral trains continued to run from nearby Waterloo station, the LNC's own terminus was never used again and after 1945 the LNC decided that reopening the railway was not financially viable and the surviving railway adjacent company office building at Westminster Bridge Road was sold. Track remained in situ at the former LNC platform at least until 1952 as evidenced by these images and the Westminster Bridge Road building remains today. Coffins were still occasionally transported by rail, including the great and the good such as Sir Winston Churchill in 1965, but in 1988 BR announced that it would no longer do so.

The air raid of 16 April 1941. This Southern Railway coach had been stabled in the platform at the Necropolis station with the results seen. The official report states, 'At Waterloo, bombs were falling all around at 10.30pm, and at 11.00pm the Station was closed and owing to HE (high explosive) on the tracks near Vauxhall all traffic to Clapham Junction was stopped. The Necropolis Station was destroyed, with four coaches burnt out (including the funeral train) and one staff injured.'

Top: The remains of a very decrepit bomb damaged Necropolis station are surveyed by the Waterloo ASM in this view taken on 14 March 1952. *LSDC1266*

Bottom left: No 1345S, one of two vehicles seemingly left abandoned at the platform, is this ex-L&SWR six wheeled guards brake carrying the markings 'Engineer's Ironworks, Wimbledon, Bridge Painters' and previously numbered D893. *LSDC1264*

Bottom right: The other vehicle was a former L&SWR four wheeled luggage van with an arc roof formerly numbered D926. *LSDC1265*

This ex-L&SWR Hearse Van numbered 1426, formerly D928, and previously used on trains from Necropolis station, is seen at Rotherhithe Road in 1950. *LSDC0688*

The imposing exterior of the Necropolis station entrance was latterly named Westminster Bridge House and can be seen today at 121 Westminster Bridge Road. *Wikipedia GNU Free Documentation License.*

New signal boxes @ Euston and Three Bridges

Euston's new signal box with its Westinghouse Power Lever Frame was opened by British Railways on 5 October 1952. It was to remain in use for almost 13 years until 27 September 1965 when, following the rebuilding of Euston station in connection with the 25kv electrification scheme, the operation of Euston passed to a new Power signal box.

The 1952 box contained a Westinghouse Style 'L' power frame, all electrically locked, consisting of 227 levers. The frame was constructed from parts originally destined for a box at Preston North. The box was manned by three signalmen on duty at any one time operating a three shift system. The new box replaced three former signal boxes, Euston No1, an 1892 L&NWR non standard design box, Euston No2, an L&NWR Type 4 design dating from 1891 and Euston No3, a 1905 built L&NWR box. The former Euston No4 box was renamed Euston Carriage Sidings, remaining in use until closure in 1965 when the 1952 box was taken out of service. One of the other casualties of the 1952 resignalling was the L&NWR brass bell which had been used from 1892-1952 to indicate the platform for all trains arriving on the east side of the station, however, it can now be seen, and possibly heard, in the Science Museum.

On 2 October 1954 Royal Scot class No 46101 *Royal Scots Grey* passes the 1952 signal box, which carried the message for weary travellers that Euston was only 300 yards further, with a lengthy arrival from the North. This Camden based 4-6-0 would later that week be transferred to Crewe North depot, only to move back to Camden the following month. *R C Riley 5552 / Transport Treasury*

Also opening in 1952 was a new signal box at Three Bridges on the London – Brighton main line which became operational in April that year. This box replaced a former LB&SCR box and had 142 levers representing the largest mechanical locking frame on the Southern Region. It would remain in service until 1983. Notwithstanding the new signal box, the colour light signalling has yet to be modernised with two separate heads denoting differing routes instead of what would later be a single head with 'feathers'. *JGS002568*

February 4 saw the inauguration of the first phase of the scheme to electrify the trans-Pennine routes from Wath-on-Dearne and Sheffield to Manchester using the 1500v DC overhead system. The initial phase has seen the electrification of the route from Wath to Dunford Bridge on the eastern side of the Woodhead tunnel.

The second phase will see the Manchester suburban services as far as Glossop and Hadfield electrified with the final phase linking up Dunford Bridge with Hadfield. Although the scheme was authorised back in 1936 the intervention of WW2 delayed matters. The main freight traffic is coal from the South Yorkshire mines which is marshalled at Wath for transport via the Woodhead route which includes the three miles of fearsome 1 in 40 grade of the Worsbrough incline. Following electrification this challenge is now surmounted with ease at some 30mph by two electric locomotives working in tandem whereas in steam days it often required two locomotives at the head of the train with two more banking, crawling up the grade at very low speeds.

A total of 65 locomotives, of which 58 will be of Bo-Bo configuration, will be provided for the scheme with 30 of these due to enter service for Phase 1. These traction units have been designated EM1 (Electric Mixed Traffic 1) latterly becoming Class 76. One locomotive has been preserved at the NRM; No 26020, later 76020, having been selected as it had been exhibited at the Festival of Britain in 1951 and had later been the locomotive that powered the opening day train through the Woodhead tunnel. An image of No 26020 also featured in *Railway Times No 4*.

EM1 No 26002 is seen outside Wath Electric Depot on 25 October 1959. *G36/1*

EM1 later Class 76 No 26047 *Diomedes* and No 26041 are seen parked up at Wath in this undated view. Diomedes was a Greek hero who participated in the Trojan War and was one of the names from classical mythology bestowed on 13 of these locomotives until removal in 1970. *GS1/70/5*

On 14 May 1978 No 26020 is under restoration at Doncaster Works for preservation at the NRM. Also in view are Deltic No 55019 and Class 37 No 37070. *BMS*

New Standard 2-6-2Ts emerge from Swindon Works

The first 20 of this new tank engine class were constructed at Swindon between April and December 1952. These were followed by 25 more produced between 1954-5. The first 10 were delivered to Tyseley, with seven going to Exmouth Junction and three to Eastleigh. The first examples were withdrawn in 1964 although two lasted in service on the Southern Region until July 1967.

This Standard design was essentially a hybrid of the LMS Ivatt class introduced in 1946 and the standard GWR boiler design used on the large prairies of the GWR and on the 56xx class tanks, although the boiler of the Standard class locomotive was slightly shorter than that used on the GWR. The use of a GWR boiler design was decided upon as there was no suitable LMS boiler design which would enable the weight restriction, for which the locomotive was designed, to be achieved.

It had been the original intention to build 63 of these locomotives but owing to rapid dieselisation the order was cut back to 45. Another result of dieselisation was that, although intended for a projected working life of 40 years, the class was extinct after less than 15 years with the shortest lived member lasting less than nine years. Although none were preserved, the '82045 Steam Locomotive Trust' is in the process of building the next member of the class. Unlike many of the current new build projects, which are aiming to recreate larger main line types, this new locomotive is intended specifically for heritage line use.

Having been completed on 30 May 1952 and looking resplendent in ex-Works condition, No 82007 is seen here at Swindon on 15 June 1952 prior to its initial allocation to Tyseley (84E) a few days later. The engine carries a 'Built Swindon' and year plate on the framing below the smokebox - see inset . Locomotives intended for the Southern Region had additional lamp irons fitted to the smokebox to carry that region's route codes. As delivered No 82007 is in the attractive lined black livery but later examples repaired at Swindon would be repainted in unlined green. _NS207841_

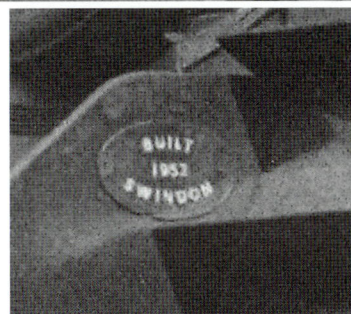

Signori Franco e Crosti

This official image dating from 1955 shows the Franco-Crosti boiler newly fitted to 9F No 92024. *MPM 999-00434*

A proposal was put forward in 1952 to try out the Franco-Crosti type of boiler on 10 of the planned new 9F 2-10-0s. These would not be constructed at Crewe Works until 3 years later in 1955 and would be numbered 92020-29. This type of boiler had been designed back in the 1930s by Attilio Franco and Dr Piero Crosti.

In essence the main difference between it and a conventional boiler was that the Franco-Crosti boiler used both exhaust steam and exhaust gases from the firebox whereas conventional feedwater heaters only used exhaust steam. Heat remaining in the exhaust gases is used to preheat the water supply for the main boiler using a secondary heat exchange mechanism known as the feedwater heater which was essentially a secondary boiler. The preheated feedwater is fed at full boiler pressure into the main boiler via clack valves. When the locomotive is fired up cold water is fed directly into the main boiler which operates normally with the exhaust gases flowing out of the main chimney via the smokebox. Once the boiler is producing steam, the exit from the smokebox into the main chimney is closed and exhaust gases instead flow through the feedwater heater and exit via the secondary chimney located towards the rear of the locomotive. Water fed into the boiler is now pre-heated in the feedwater heater and enters the boiler at a higher temperature than normal.

The railways of Belgium, Italy and West Germany all had locomotives fitted with these boilers and based on their experiences BR decided not to be left out and as mentioned above signalled their intention in 1952 to try out this type of boiler. In the BR design the standard chimney was still used for exhausting smoke from the firebox but the steam chimney was located on the right-hand side, just forward of the firebox, doing nothing for the aesthetics of the design. When running at speed the updraft on the smoke chimney was all that was needed however, where more power was needed steam would be sent to the blastpipe to draw more

air through the firebox. The experiment was not considered a success as less improvement than had been expected was forthcoming. There was also a maintenance problem owing to the acidic flue gases condensing in the feedwater heater leading to corrosion. All ten of the 9Fs so fitted were converted back to a more standard form within a few years although the preheaters remained in place although blanked off.

No 92023 was to be put through its paces at the Rugby testing plant between 7 June – 28 September 1955 followed by controlled road tests between Carlisle and Hurlford. Dr. Crosti visited the site once in July and three times in August. To capture the offset exhaust fumes emanating from the unconventional position of the chimney, the test plant chimney flue had to be repositioned and enlarged at the base. The hoped for economy in fuel consumption did not result from use of this type of boiler and no further 9Fs were so fitted. This apparently did not please Dr. Crosti's financial backers who contested the Rugby results for it had been hoped to sell many more boilers of this type to BR. None other than the great Andre Chapelon was brought in to arbitrate and found in favour of Rugby. (See article on the Rugby Testing Station in *Railway Times No 1*).

Not to be outdone the ever resourceful Bulleid, whilst working for CIE, converted the 1907 2-6-0 locomotive No 356 to carry a Franco-Crosti boiler similar to the original Italian designs. The locomotive as originally converted proved to be a poor steamer and a forced draught fan was later added powered, in a somewhat Heath Robinson fashion, by a bus engine carried on a wagon behind the tender. In view of the Republic's lack of coal reserves experiments were conducted with No 356 between 1952-1954 using peat as a fuel which was fed by screw from an enlarged tender. This was the predecessor to the infamous CC1 'Turf Burner'.

Wellingborough shed, where they were universally disliked by fitters and train crew alike, was home to the Crosti boilered 9Fs including No 92026 seen here in 1955. *NS207696a / Transport Treasury*

Perhaps best to draw a veil over this hideous apparition that was the CIE's Class K3 No 356 after it had been fitted with the Franco-Crosti 'gubbins'. It is seen here at Inchicore in June 1954. *NS208903 / Transport Treasury*

Royalty goes by train

A selection of journeys over BR metals made by royalty during 1952 included –

Euston to Edinburgh and return on 25-27 June conveying Queen Elizabeth

Euston to Ballater (for Balmoral) on 7-8 August conveying Queen Elizabeth, the Duke of Edinburgh, Prince Charles and Princess Anne

Paddington to Llandrindod Wells and Shrewsbury on 22-24 October conveying Queen Elizabeth and the Duke of Edinburgh

King's Cross to Newcastle on 26-27 November conveying Princess Margaret.

On Southern Region routes two sample journeys are featured opposite, the first being when the yet uncrowned Queen visited Devon between 1-3 July 1952. The Royal Train left from Paddington and journeyed to Newton Abbot where during a 6½ hour visit Her Majesty visited the town, the first provincial town to be so honoured since she acceded to the throne in February that year and apparently the first British monarch to grace Newton Abbot since Charles I in 1625.

She then toured the Royal Agricultural Showground at nearby Stover. Whilst she was passing through the crowds a young girl pressed her way to the front remarking that the Queen "wasn't wearing her crown". The Queen replied with a smile "No, I have left it at home today" although of course she would not be crowned until June the following year. Her train returned to London via Exeter and Salisbury to Waterloo headed by Merchant Navy Pacific No 35025 *Brocklebank Line*.

The second occasion was when the Queen visited HMS Daedalus, alighting at Fort Brockhurst station on the Fareham – Gosport branch where Her Majesty was greeted by the Mayor of Gosport and then driven to HMS Daedalus. At the base she inspected troops and watched a fly past of no less than 327 aircraft from 11 squadrons of the Royal Naval Air Command. The train departed at 10am from Waterloo on 21 November 1952 and consisted of five Pullman coaches hauled by West Country Pacific No 34011 *Tavistock*. The return rail journey commenced at Fareham. HMS Daedalus formally closed in 1996.

Observed by two walkers plus their canine companion and typical of the Royal journeys made across the country in 1952 is this view of the Royal Train at Hillmorton near Rugby appropriately headed by immaculate Royal Scot class No 46126 suitably named *Royal Army Service Corps*. MC30527

Merchant Navy No 35025 seen here at Vauxhall with the traditional headcode disc pattern of one placed centrally at the top of the smokebox and three above the bufferbeam. *LSDC1391*

In different weather conditions, West Country No 34011 passes Vauxhall in the gloom on the outward journey with the Royal Train to Fort Brockhurst. *LSDC1527*

Time's up for the 'Middy'

The Mid Suffolk Light Railway's line from Haughley – Laxfield closed to passengers on 28 July 1952.

The Mid-Suffolk Light Railway (MSLR) was intended to open up an agricultural area of central Suffolk taking advantage of the reduced construction costs permitted by the 1896 Light Railways Act. Although it was launched with considerable enthusiasm by local interests who planned to construct a 50 mile network, the actual share subscription was poor and only 19 miles from the main line at Haughley to Laxfield was opened to goods traffic in 1904. Passing into receivership in 1906, a mile and a half extension to Goram's Mill at Cratfield was opened the same year but this was short lived, closing in 1912. A passenger service began in 1908 and the line was absorbed by the LNER in 1924. During WW2 the number of passenger journeys was reduced from four to two each way daily and the end finally came on 28 July when all services ceased. In the early 1990s a group of enthusiasts formed a company to recreate the Middy Line at the site of Brockford & Wetheringsett station and they currently run regular steam services.

Opposite top: In this pre-nationalisation image LNER Class J15 No 5470 waits in the bay platform at Haughley with the Laxfield train. From November 1939 Laxfield trains used this LNER platform at Haughley, the separate MSLR station being retained for goods purposes only. *E1378*

Opposite bottom: No 65467 is handling a Haughley to Laxfield mixed train and is seen at the intermediate station of Worlingworth. These 0-6-0s of the J15 class were the usual motive power on this route. *E3323*

Above: On 24 September 1949 Class J15 No 65459 carrying the new British Railways lettering on its tender, applied three months before in June 1949, prepares to leave the rudimentary passenger terminus at Laxfield with its train consisting of six wheel brake third E62331, six wheel comp E63404 and a four wheel goods brake. *OTABD0775*

Opposite top: No 65447 which has charge of a mixed working, the 11:8am from Haughley to Laxfield, crosses No 65361 with a Laxfield to Haughley goods working at Kenton in 1951. *E012*

Opposite bottom: A sad occasion as No 65447, suitably adorned with a wreath, takes water at Laxfield on the final day of passenger services on the Mid Suffolk Light. *E008*

This page, top: Unheard of crowds throng the platform at Laxfield to see the arrival of No 65447 on the last day of passenger services on 28 July. *E048*

This page, bottom: No 65388 has the melancholy task of hauling one of the demolition trains and is seen here at Laxfield Mill amongst the grass grown tracks. This extension, 1½ miles from Laxfield station, opened in 1906 to serve the wind powered Goram's Mill at Cratfield. *E066*

Too wise you are.......

'**2 Ys UR, 2 Ys UB, ICUR 2Ys 4Me.**' If you recognise that phrase then you probably passed around an autograph book at the end of your schooldays during the 1950s or 1960s as it was often used to preface your school friends' signatures. The three types of Class Y 0-4-0s produced by Sentinel are featured here, 1952 proving to be the last year of operation for the Y10.

The LNER Class Y1 was a numerically small class of 0-4-0 geared locomotive built by Sentinel and introduced in 1925. Passing into BR ownership in 1948 they were numbered 68130-68153 although Nos. 68134 and 68135 were withdrawn in the year of nationalisation. The superheated vertical water tube boiler was similar to those used in Sentinel steam wagons. There were variations within the class regarding boiler size and fuel capacity and these were denoted by sub-classes Y1/1 to Y1/4. The locomotives had poppet valves and the advantage of the water-tube boiler was that it enabled steam to be raised much more quickly than with a conventional fire tube boiler. There was a total of 7 Y1/1, 15 Y1/2, 1 Y1/3 and 1 Y1/4. Class Y1/2 number 68153 is preserved on the Middleton Railway in Leeds, having been the last survivor in BR service when withdrawn from Darlington in June 1961.

The Y3 class had a two-speed gear box, whereas class Y1 had only a single speed, otherwise the classes were identical in appearance. A total of 32 Y3s were purchased from Sentinel between 1927 and 1931, all Y3s having the larger boiler as fitted to the Y1/2 variants. Over 40 Sentinel locomotives and railcars exist today at various locations around the world and in addition to the Y1/2 mentioned above, a non BR shunter of Y3 design also survives in the UK at the Buckinghamshire Railway Centre.

The other Class Y that Sentinel produced for the LNER was the Y10 which was a class of just two introduced in 1930. Numbered 8403 and 8404 they were later re-numbered 8186 and 8187 with both passing into BR ownership, although No 8187 was withdrawn almost immediately. No 8186 was allocated the BR number 68186 but this was never actually carried before withdrawal came in 1952. They were intended for service on the Wisbech & Upwell tramway and were fitted with cowcatchers and sideplates with a cab at each end. They were not however a great success on the tramway and both were subsequently moved to Yarmouth in May 1931 to work the quayside lines.

Three examples of these Y Class 0-4-0s are illustrated below.

Opposite: With lumps of coal perched rather precariously Class Y1 No E8130, seen in Lowestoft at the junction of Station Square and Waveney Road, is about to cross over to the Fish Quay under the control of a flagman. Walker Regis, a well known emporium in the town, is prominently advertised together with Norwich Union. Following Nationalisation it was given the number 68130 but this was never applied and by the early 1950s it had become No 37 in the Departmental series. *NS207553 / Transport Treasury*

Above: From July 1951 Wrexham Rhosddu shed had been the base of this Class Y3, No 68162. It would be withdrawn from 6E in July 1960. *NS207553 / Transport Treasury*

Right: Yarmouth Vauxhall is the location of this image of Class Y10 No 68186. This Sentinel just made it into 1952, being withdrawn in February. *NS207549 / Transport Treasury*

Above: The loco crew are struggling with the 52 foot turntable at Stratford Old Town on 8 April 1958 during the line's freight only existence. Fowler 4F 0-6-0 No 44434 is their mount whilst in the background a line of wagons contains an interesting contingent of army vehicles. The shed here had officially closed on 22 July the previous year and was soon demolished, leaving only the water tank and the turntable which is believed to have remained operational into the mid-1960s. *RCR11557 / Transport Treasury*

Slow, Middling & Jolty - the passing of the S&MJR

At nationalisation the management of the S&MJR line was divided organisationally between the LMR and the WR with the boundary being located at Fenny Compton. Local passenger services in such a sparsely populated rural area had long been subject to more attractive bus operations and BR sought to reduce the heavy losses that were being incurred. The first to go under the new nationalised ownership was the short northward spur from Byfield leading to the Great Central line at Woodford Halse, the passenger service ceasing from 31 May 1948. The extension from Stratford to Broom Junction had previously closed in June 1947 with the short lived passenger service from Towcester to Olney only lasting four months before premature withdrawal came in 1893.

With the number of passenger bookings between Blisworth and Stratford-upon-Avon down to just single figures daily by the early 1950s closure became inevitable and, following closure of the Banbury to Towcester service in July 1951, closure of the route from Blisworth to Stratford duly occurred from 5 April 1952 although freight traffic continued until 1965. However the core S&MJR route was identified as having a strategic long-distance freight carrying potential such that from 4 June 1951 four through freight trains, often headed by Black 5s and 9Fs, each way daily were routed from the Great Central line at Woodford to Broom South Junction, continuing through Ashchurch to South Wales carrying iron ore. The line between Ravenstone Wood Junction and Towcester subsequently closed to freight in June 1958 and during 1964/5 all the residual local freight traffic was dispensed with and the through trains diverted away. The only remaining operational part of the former S&MJR system that remained was the connection from Fenny Compton to Kineton MOD establishment, the stub of the branch being transferred to MOD ownership in July 1971.

Opposite bottom: Taken from the top of the water tower, the shed lies just off the picture to the right whilst the background is dominated by the rail connected grain silos which were situated close to the riverside location of Lucy's grain mill. These silos, one of several constructed across the country by the Government during WW2, were demolished in the 1970s. *RCR11554 / Transport Treasury*

Above: 1957 is the year recorded by the photographer for this shot of the station and signal box at Old Town. In view of the number of people milling about it was probably taken during a visit by enthusiasts, possibly that of 11 May that year organised by the Gloucestershire Railway Society which saw Compound 4-4-0 No 40930 power the train for at least part of the journey. *NS203585 / Transport Treasury*

This page, top: Class 4F No 44567 waits at Blisworth on the last day of services over the S&MJR on 5 April 1952 with the departure for Stratford. The 0-6-0 sports a 21D shedplate denoting Stratford-upon-Avon depot which, at this date, had an allocation of eight of these 4Fs plus 3 Johnson 3Fs and a solitary Deeley 3F. *OTA0406-102*

This page, centre: A scene of grass grown dereliction looking west at Towcester, formerly a junction of four routes but today the site of a Tesco supermarket. Of note are the signal box which contained a 50 lever frame and the unusual signal post on the right. *OTA32470*

This page, bottom: A May 1961 view of the eastern junction at Towcester with the line to Blisworth heading off to the left whilst the freight only line to Olney is seen on the right underneath the road bridge which has required some support. This latter route has an unfortunate claim to fame in that passenger services between Olney and Towcester only lasted for four months in 1892/3 before premature withdrawal owing to lack of traffic. *NS203589 / Transport Treasury*

Opposite top: This is Byfield on the S&MJR with the RCTS 'Grafton' railtour of 9 August 1959 powered by Black Five No 45091 which operated the leg from Blisworth to Byfield. Here it ran round to take the tour on to Woodford Halse, Brackley, Verney Junction and Banbury Merton Street. *JH944*

Opposite bottom: 17 April 1956 saw the photographer at Fenny Compton where he took this image showing the single track of the S&MJR route on the left and the double track of the former GWR main line to Leamington Spa on the right. *RCR7039 / Transport Treasury*

The contrasting livery of these two Pacifics seen at Edinburgh's Haymarket shed allows us to date this view to between August 1951 and April 1952. A3 Pacific No 60035 *Windsor Lad,* seen on the turntable, received its Brunswick green livery in August 1951 whilst the blue livery applied to A4 Pacific No 60011 *Empire of India,* seen in the background, was replaced with green in April 1952.

No 60159 *Bonnie Dundee* is also seen at Haymarket in this view dating from 1951/52 before the blue livery applied to this A1 Pacific from new in 1949 was replaced by green in December 1952. Like all members of its class it carried from new the rather austere unlipped chimney which was later replaced with a more stylish lipped version. It only saw service for a little over 14 years before being scrapped at Inverurie Works in January 1964. Incidentally 'Bonnie Dundee' was the title of a poem and a song written by Sir Walter Scott in 1825 in honour of John Graham, 7th Laird of Claverhouse, who was created Viscount Dundee in November 1688. Following this, in 1689, he led an uprising in which he died, thus in the process becoming a Jacobite hero. *Both David P Williams colour archive*

One of the workhorses of the NER was this Class P3, later J27, No 65865 seen here working hard on the 1 in 125 incline of Stockton bank in 1952. It was a regular performer in this area, remaining at work until withdrawal as late as February 1967. Of interest in the background is the signal post with lower quadrant ex-NER signals still in evidence.

Also on Stockton Bank near Norton South signal box is ex-NER Class R, later class D20, No 62347 with a local service for Sunderland. By 1952, the date of this image, many of this class of 4-4-0 had been withdrawn with the survivors eking out an existence on local trains and excursion services. This Northallerton based locomotive would be withdrawn a couple of years later in November 1954 after a working life of close on 55 years. *Both David P Williams colour archive*

Slinging their hook - Isle of Portland closure

In his famous series of Wessex novels author Thomas Hardy rechristened several locations, one of which was the Isle of Portland which became his 'Isle of Slingers'. By 1952 BR had decided that they had had enough of providing unremunerative passenger services to the terminus at Easton and to the intermediate station of Portland and so 'slung their hook' and withdrew from this market although continuing to provide goods services as far as Portland station until 1965.

The writing was on the wall – well on the poster at least – as this view of the closure notice for the Weymouth – Portland service posted at Waterloo indicates. *LSDC1292*

This page: The scenic attractions of this cliff hugging route are apparent in this view of Class O2 tanks Nos. 30179 and 30197 rounding the curve with the 3:17pm service from the terminus at Easton. Yeolands Bridge spans the track which enters a cutting on the approach to Easton station. *LSDC1259*

Opposite top: The Easton terminus with its attractive stone built station and single road engine shed, outside of which is the water tower, is seen here in 1952 whilst an unidentified tank locomotive stands by the signal box. *LSDC1257*

Opposite bottom: With evidence of the extensive quarrying that occurred across Portland seen on the right, the two O2 tanks seen previously head an earlier departure from Easton at 1:53pm. *LSDC1254*

BRITISH RAILWAYS

WITHDRAWAL OF PASSENGER TRAIN SERVICE
FROM
WEYMOUTH-PORTLAND-EASTON LINE

On and from Monday, 3rd March, 1952, the passenger train service between WEYMOUTH (MELCOMBE REGIS), PORTLAND and EASTON will be withdrawn and the stations on the line closed to passengers.

Omnibus services covering the area are provided by the Southern National Omnibus Company Ltd. and these will be augmented as necessary on and from 3rd March, 1952.

Further information may be obtained from, the Stationmaster, WEYMOUTH (telephone Weymouth 850), the Stationmaster, PORTLAND (telephone Portland 2130) or from the District Traffic Superintendent, British Railways, SOUTHAMPTON (telephone Southampton 3838, Ext. 686). Enquiries regarding omnibus arrangements should be made to The Southern National Omnibus Company Ltd., 5 Royal Terrace, WEYMOUTH (telephone Weymouth 645).

PARCELS AND FREIGHT TRAIN TRAFFIC WILL CONTINUE TO BE HANDLED

Passing the signal box the crew of Class O2 tank No 30230 take the opportunity for a word with the signalman. Prior to its transfer to Feltham depot No 30230 was based at Dorchester shed from October 1951 until March 1952, helping to date this image. *BMS*

A closer view of the Easton signal box seen in the early 1950s. It remained open after passenger closure in 1952 until April 1955 in order to handle the daily goods service. *LOSA20874*

Putting their backs into it - 1952 style

A fascinating scene of engineering works, involving some back breaking shovelling by the looks of things, is revealed in this 8 July 1952 view taken at Blaby sidings near Wigston showing Johnson 3F No43326 in attendance. *HG0350*

Getting away from a Penrith stop on an unrecorded date in 1952 is an up service headed by nearly new Clan Pacific No 72004 *Clan Macdonald*. *NS209207 / Transport Treasury*

Following the recent introduction of the Britannia Pacifics, the new 6P5F two cylinder Pacifics named after Scottish clans entered service at the end of December last year with the final one of the 10 ordered being released from Crewe Works in March this year.

As their naming implies all are destined for the Scottish Region. Whilst the chassis is similar to the Britannia class the boiler is smaller, enabling the weight to be kept down, giving the locomotives good route availability. It is anticipated that they will be used on passenger and fast freight workings and special provision has been made for the fitting of tablet catching apparatus for working on the lengthy stretches of single track encountered on main lines north of the border.

BR	Date to service	Name	BR	Date to service	Name
72000	29/12/1951	Clan Buchanan	72005	02/1952	Clan Macgregor
72001	29/12/1951	Clan Cameron	72006	27/2/1952	Clan Mackenzie
72002	14/1/1952	Clan Campbell	72007	02/1952	Clan Mackintosh
72003	19/1/1952	Clan Fraser	72008	02/1952	Clan Macleod
72004	02/2/1952	Clan Macdonald	72009	03/1952	Clan Stewart

Opposite top: No 72003 *Clan Fraser* heads a Glasgow to Manchester express at Penrith in June 1952. *FH965 / Transport Treasury*

Opposite bottom: The former L&NWR Manchester Exchange station witnesses the steamy departure of the first of the class, No 72000 *Clan Buchanan*, in this undated view. Closing to passengers in May 1969, the station continued to host newspaper trains until the 1980s after which it continued to operate for some years as a car park. *NS203832 / Transport Treasury*

A4 Pacific No 60011 *Empire of India* heads 'The Elizabethan' through Peterborough North on 5 September 1953. ('The Elizabethan' was the new name for the former 'Capitals Limited' express renamed to celebrate the coronation of Queen Elizabeth II and running non stop between King's Cross and Edinburgh Waverley.) *ES1120 / Transport Treasury*

In excellent external condition No 2920 *Saint David*, the last of the 'Saints' to be withdrawn, is seen at Gloucester in 1952. *NS201079 / Transport Treasury*

Amongst the items it is intended to feature the following –

The 1953 Transport Act

The Great Storm - flooding of east coast and SR lines

Bibby Line axle fracture at Crewkerne

'Starlight Specials' - London - Scotland overnight

More closures on the Isle of Wight

New locomotives for BR

The railways celebrate the Coronation

and lots more!